Operation Code

Coding at the Zoo

Scratch 3.0

By Colleen van Lent and Kristin Fontichiaro

CHERRY LAKE PRESS

Published in the United States of America by Cherry Lake Publishing
Ann Arbor, Michigan
www.cherrylakepublishing.com

Content Adviser: Series Adviser: Kristin Fontichiaro
Reading Adviser: Marla Conn, MS, Ed., Literacy specialist, Read-Ability, Inc.

Image Credits: Various images throughout courtesy of Scratch

Library of Congress Cataloging-in-Publication Data

Names: van Lent, Colleen, author. | Fontichiaro, Kristin, author.
Title: Coding at the zoo / by Colleen van Lent and Kristin Fontichiaro.
Description: Ann Arbor, Michigan : Cherry Lake Publishing, 2020 | Series: Operation code | Includes index. | Audience: Grades 2-3.
Identifiers: LCCN 2019035828 (print) | LCCN 2019035829 (ebook) | ISBN 9781534159242 (hardcover) | ISBN 9781534161542 (paperback) | ISBN 9781534160392 (pdf) | ISBN 9781534162693 (ebook)
Subjects: LCSH: Scratch (Computer program language)–Juvenile literature. | Computer programming–Juvenile literature. | Zoos–Juvenile literature.
Classification: LCC QA76.73.S345 V359 2020 (print) | LCC QA76.73.S345 (ebook) | DDC 005.13/3–dc23
LC record available at https://lccn.loc.gov/2019035828
LC ebook record available at https://lccn.loc.gov/2019035829

Cherry Lake Publishing would like to acknowledge the work of the Partnership for 21st Century Learning, a Network of Battelle for Kids.
Please visit *http://www.battelleforkids.org/networks/p21* for more information.

Printed in the United States of America
Corporate Graphics

NOTE TO READERS: Use this book to practice your Scratch 3 coding skills. If you have never used Scratch before, ask a parent, teacher, or librarian to help you set up an account at *https://scratch.mit.edu*. Read the tutorials on the website to learn how Scratch works. Then you will be ready for the activities in this book! You will practice using variables, if/then statements, copying code to other sprites, using effects to change a sprite's look, and more! Find all the starter and final programs at *https://scratch.mit.edu/users/CherryLakeCoding*.

Table of Contents

Move Baby Bear

It is spring. Time for Baby Bear to wake up and move around. We want to use Scratch, but we bears have a hard time **dragging and dropping** code. (No fingers!)

I got things started at *https://scratch.mit.edu/projects/312040594*. Can you help me with the rest?

Pro Tip!

Scratch lets you see other people's projects and copy them to make them your own. My code contains the **sprites**, background, and sounds we will use in this book.

Ways to Make a Sprite Move

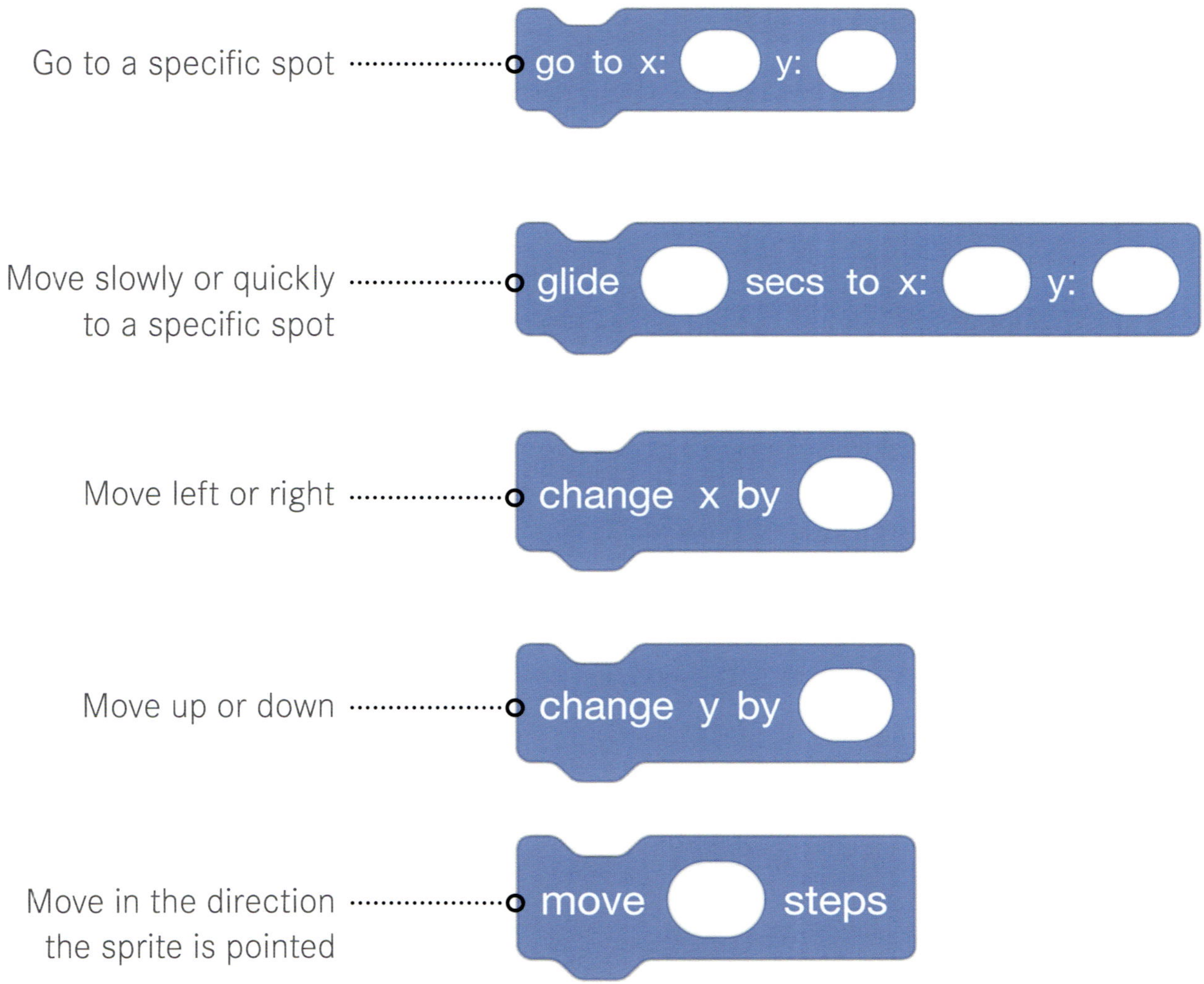

Motion Blocks

Scratch has a lot of ways to move a sprite. You can use *move, go to, change x,* or *change y*. You can find these blocks in the Motion **category**.

Let's move Baby Bear by dragging this Motion block into his code:

Pro Tip!

Motion blocks can move in many different ways. Certain blocks move sprites right, left, up, down, or diagonally. Remember, these blocks move in the direction the sprite is facing.

```
when [right arrow ▼] key pressed
move (10) steps
```

```
when [left arrow ▼] key pressed
move (10) steps
```

Using the Arrow Keys to Move

Baby Bear has been sleeping for months. He wants to move! Let's move Baby Bear with the arrow keys on our keyboard.

To do this, we need to go to the Events category. Add the code seen on page 8.

Uh-oh! Baby Bear always moves left, even when we press the right arrow key! Let's fix that.

Pro Tip!

We used the **default** 

Events block. We clicked on the white down triangle in the block to find the *right arrow* and *left arrow* choices.

Moves your sprite to look up

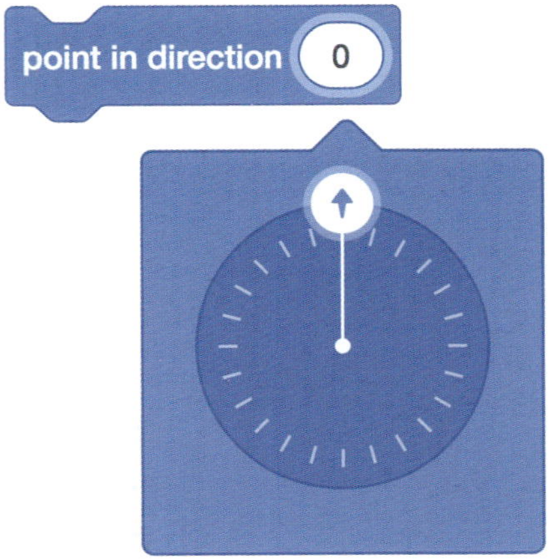

Moves your sprite to look right

Moves your sprite to look down

Moves your sprite to look left

The Right Direction

We need to tell Baby Bear which direction to face. To do that, we use this block:

point in direction (90)

You can find this in the Motion category.

Scratch uses numbers for directions. This chart will help you match the number to the direction you want.

Pro Tip!

Scratch can tell you which number to use! Click on the white circle inside *point in direction*. A circle will pop up. Move the circle's arrow to the direction you want your sprite to face. Scratch will update the number.

```
when [right arrow] key pressed
point in direction (90)
move (10) steps
```

```
when [left arrow] key pressed
point in direction (-90)
move (10) steps
```

Putting It Together

Now that we understand how the `point in direction (90)` Motion block works, let's add it to the code.

When you are done, test your code. Is Baby Bear looking in the direction he's walking?

Pro Tip!

When you have more than one **code segment**, you can place them anywhere on the screen.

```
when [right arrow v] key pressed
set rotation style [left-right v]
point in direction (90)
move (10) steps
```

```
when [left arrow v] key pressed
set rotation style [left-right v]
point in direction (-90)
move (10) steps
```

What Can Go Wrong?

Sometimes the **point in direction (90)** block doesn't do exactly what you want. What might happen? Baby Bear may end up upside down! To make sure this doesn't happen, add this block:

set rotation style [left-right ▾]

This Motion block will tell Scratch to flip Baby Bear left and right, not upside down.

Pro Tip!

Use the *left-right* option in the *set rotation style* **Motion** block to make the sprite only rotate left and right. Use the *all around* option for up and down movement.

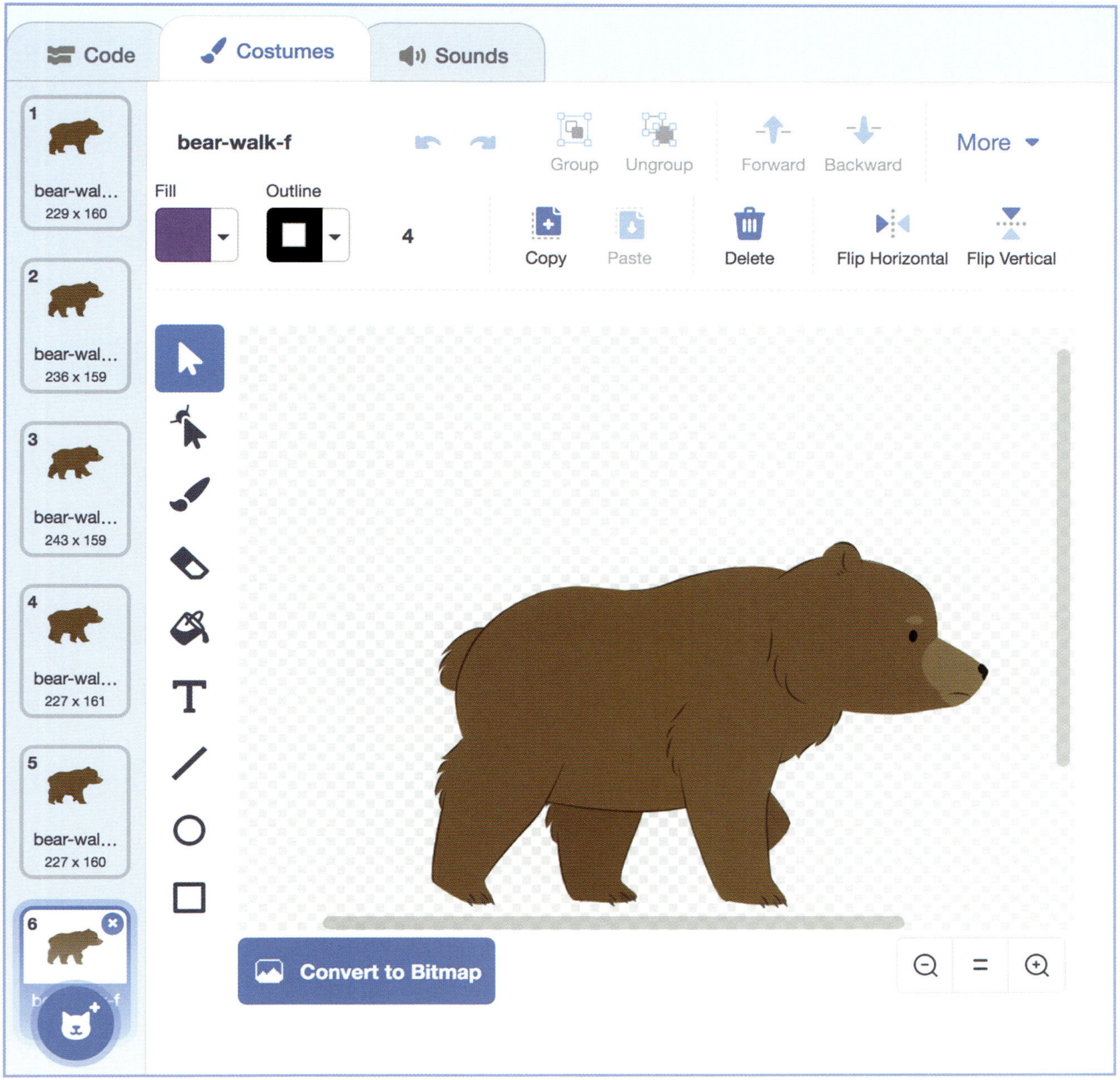
Code
Costumes
Sounds
1
bear-wal...
229 x 160
2
bear-wal...
236 x 159
3
bear-wal...
243 x 159
4
bear-wal...
227 x 161
5
bear-wal...
227 x 160
6
bear-walk-f
Group
Ungroup
Forward
Backward
More
Fill
Outline
4
Copy
Paste
Delete
Flip Horizontal
Flip Vertical
Convert to Bitmap

Move Your Feet!

Baby Bear is moving, but it doesn't look like he's walking.

Click on the **Costumes** tab. He has eight looks. Each costume puts his feet in a different position. When you use the `next costume` block, Scratch will change to the next look. You can find this block in the Looks category.

Add this block to the end of your code. Now Baby Bear will look like he is walking!

Pro Tip!

Only some sprites have different walking costumes. Click on the cat **icon** in Costumes and search "walking." You'll see a list of sprites that can look like their feet are moving.

```
when this sprite clicked
say (Hello!) for (4) seconds
```

Hello!

Let's teach Baby Bear how to say "Hello!" Which event should we use? Let's try this Events block:

```
when this sprite clicked
```

Then add this Looks block to the code:

```
say ( ) for ( ) secs
```

You can change "Hello!" to other words and phrases, too.

Pro Tip!

There are lots of ways that sprites can say or think things. You can find them in the **Looks** category.

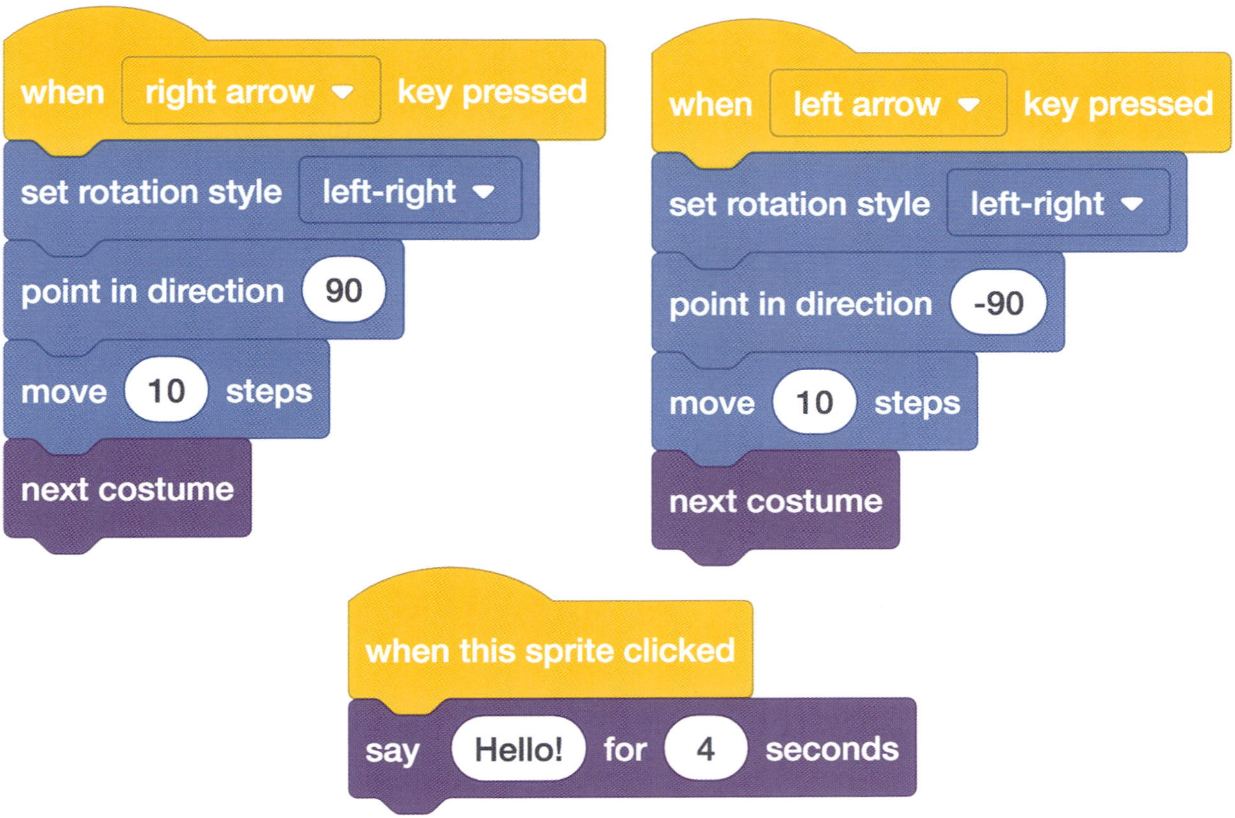
when right arrow key pressed
set rotation style left-right
point in direction 90
move 10 steps
next costume
when left arrow key pressed
set rotation style left-right
point in direction -90
move 10 steps
next costume
when this sprite clicked
say Hello! for 4 seconds

Baby Bear Is Ready to Play!

Thank you for helping Baby Bear wake up from **hibernation**! Check out his final code. What other things can you help him do?

Here are some ideas:

- Use Sound to help Baby Bear grunt.
- Add a Big Sister for Baby Bear.
- Change what Baby Bear says.

What other ideas do you have?

Pro Tip!

If you would like to see our final code, go to *https://scratch.mit.edu/projects/311734176*.

Glossary

category (KAT-uh-gor-ee) a group of things that have something in common

code segment (KODE SEG-muhnt) a collection of Scratch blocks that are attached together

costumes (KAHS-toomz) a sprite's different poses in Scratch

default (DEE-fawlt) the choice Scratch makes for you in advance that you can change

dragging and dropping (DRAG-ing and DRAHP-ing) clicking and holding down the mouse while moving a block

hibernation (hye-bur-NAY-shuhn) sleeping through the winter to save energy

icon (EYE-kahn) a little picture that you can click on

sprites (SPRYTS) characters or objects in Scratch

Find Out More

Books

LEAD Project. *Super Scratch Programming Adventure!* San Francisco, California: No Starch Press, 2019.

Lovett, Amber. *Coding with Blockly.* Ann Arbor, Michigan: Cherry Lake Publishing, 2017.

Websites

Scratch
http://scratch.mit.edu
Build your Scratch code online at this site.

Scratch Wiki
https://en.scratch-wiki.info
If you get stuck, ask an adult to help you look on this site for advice.

Index

About the Authors

Colleen van Lent teaches coding and Web design at the University of Michigan School of Information. She has three cool kids and a dog named Bacon. She wishes she could touch her toes.

Kristin Fontichiaro teaches at the University of Michigan School of Information. She likes working with kids on creative projects from coding to sewing to junk box inventions. She has written or edited almost 100 books for kids.